TERROR IN THE TROPICS: THE ARMY ANTS

by
Tom Lisker

A

cpi

Book

RSVP
RAINTREE
STECK-VAUGHN
PUBLISHERS
The Steck-Vaughn Company

First Steck-Vaughn Edition 1992

Art and Photo Credits

Cover photo and photos on pages 7 and 44, M.P.L. Fogden/Bruce Coleman.
Illustrations on pages 9, 11, 13, 22, 23, 26, 29, and 31, Jeffrey Gatrall.
Photos on pages 15, 20, 35, 37, 39, and 43, Dr. Howard Topoff.
Photos on pages 16 and 41, Oxford Scientific Films/Bruce Coleman.
Photo on page 47, Jane Burton/Bruce Coleman.
All photo research for this book was provided by Sherry Olan.
Every effort has been made to trace the ownership of all copyrighted material in this book and to obtain permission for its use.

Library of Congress Number: 77-10765

Library of Congress Cataloging in Publication Data

Lisker, Tom, 1928-
    Terror in the tropics: the army ants.

    SUMMARY: An introduction to army ants, their habits, homes, the social aspects of their lives, and how they affect human beings.
    1. Army ants—Juvenile literature. [1. Army ants. 2. Ants] I. Title.
QL568.F7L58   595.7'96   77-10765

ISBN 0-8172-1060-1 hardcover library binding

ISBN 0-8114-6866-6 softcover binding

18 19 20 21 22 23 24 25   00 99

CONTENTS

Chapter 1

THE KILLER ARMY

At dawn, the scouts move out from the base camp. They move, almost like machines, in ones, twos, and threes. Then they search the area silently and nervously. Unbelievably, these scouts can't see or hear. None of them can speak, either. Nevertheless, they move about, feeling their way, tapping and exploring.

Soon they come upon an unsuspecting victim. Noting its location, they quickly return to the main group to report their discovery—not with sounds, but with a ritual "kiss." This "kiss" is a call for help. The enemy has been sighted and it's large. More powerful soldiers are now needed to finish the job.

These soldiers are bloodthirsty hunters with long, hooked jaws. Once they have bitten their enemy—a large animal—they will not let go until they have torn out a piece of its flesh.

The scouts have left a trail that leads the soldiers to the prey. Fortunately for the victim, the killing is over quickly. Death comes before the body is brutally torn apart. Once dead, the animal is dragged back to the swarming army that hungrily awaits its food.

Sometimes the battle may rage for hours if the trapped animal tries to escape. But the doomed animal only finds itself running madly into another raiding party, and still another. Finally, all energy gone, the victim gives up all hope of escape. No one and nothing escapes these barbaric hunters. Every living creature in their path—from the tiniest insects to giant pythons and large horses—are reduced to bare skeletons. Even human beings are not safe from attack. Only those who fly have a chance against this powerful enemy. For these hunters are the mysterious and terrifying army ants.

The army ants are, in every sense, a tight military unit. They form field stations where scouts and soldiers prepare for an attack.

One of the frightening things is that army ants give very little warning of their approach. The only sound they make is a strange, rustling noise. That's the sound of their hard bodies crawling up the walls of a home. Sometimes they can be heard rushing along the ground. Sometimes the frightened cries of the animals outside may warn of a raid. And sometimes the arrival of a hungry, moving army is announced by flocks of screaming jungle "ant birds." *Sometimes!* Most often, the ants sneak up and attack without warning. Army ants are the terror of the tropics.

The "small" armies travel with only a few thousand ants. Larger armies travel with more than a million! *Some army colonies have been known to contain over 30 million ants.* Their food supply must be large enough to satisfy all of the army's members. That's a terrifying thought.

Army ants have no real home. They move about, setting up camp from place to place. They get little rest. Their lives are an almost continuous search for more and more food to feed their growing, hungry army. These killer soldiers move in on their prey in columns 65 feet wide!

What makes them so bloodthirsty? What makes them wage this kind of war? How can they be stopped? Where will they strike next? Who will their victim be?

The answers to these questions have been a mystery to people for hundreds of years. Luckily, each year, ants become less of a mystery as experts gather more information about them.

Even large animals the size of horses have been attacked by the tiny soldiers.

Chapter 2

MYTHS AND LEGENDS - ARE THEY TRUE?

Throughout history, many writers have stated that ants have fixed workdays and holidays. Some argue that ants rest on the first day of each month. Other writers are sure it is the ninth day. The fact is, no one really knows when, or even if, ants rest at all.

Other reports say that laziness in the ant world is an awful crime. It is also written that since ants work so hard most of the time, they

Some people believe ants take regular holidays.

hold regular market days to catch up with the news of their friends. Most writers describe these ants in human terms. But are they as intelligent as man? Do they really live in "humanlike" societies?

Robert Frost, the famous American poet, described the burial of an ant in one of his poems. The Queen ant was notified of the death of one of her soldiers. She informed the proper department in the ant colony. And they took care of the burial. It is said that ants, like people, honor their dead. They place their dead in "coffins" before they bury them in the ant "cemetery."

These, of course, are all myths that have been told over the years about ants. But like many myths, they are based on half-truths. Ants really *do* carry away their dead. They are carried not to a cemetery but to a refuse pile away from the living area. This is done to protect the health of the ant colony. Does this prove the intelligence of ants? Probably not. It is *instinct*—an unknown force within the animal—at work, not intelligent thought.

This "burial instinct" is set in motion by certain chemicals released by the dead ant. To prove this is true, experimenters placed these "chemicals of the dead" on some living ants. Sure enough, their friends and neighbors carried the live ants off and threw them on the refuse pile. It didn't matter to the others that these ants were alive—going about their work right before their "eyes."

Do ants really hold funerals for their dead?

13

The poor live creatures refused to remain with the dead. They dragged themselves from the heap and returned to the nest. Eventually, the chemicals wore off and they continued their work. Surely they didn't appreciate the help given by their friendly "undertakers!"

Perhaps the greatest legends of all center around the Queen. Her title alone says that she must be the most honored member of the group. After all, isn't she "the mother" of her entire colony? Isn't she cared for by attentive "maids-in-waiting" who feed, bathe, and even carry her from place to place? But once again, this is an example of people trying to make ants seem human.

Queens are certainly given special care. But this attention seems also to come from a chemical action. Biologists have successfully placed "Queen chemicals" on small objects. When the objects were placed in the same ant colony, the workers treated them just as if they were Queens! They clustered about them, licked them, and carried them blindly from place to place. This was not loyalty. It was instinct.

Another myth has to do with ant "nurses." The ant eggs are well cared for by the "nurses"

The huge Queen of the army ants sends constant chemical
signals to the members of her army.

but not because they will soon be future citizens
of the colony. That would be looking at ant behavior in human terms. The "nurses" look after
ant eggs because the eggs have certain chemical

The army ant Queen, her body swollen with eggs, is cared for by
a worker ant.

smells and tastes. In some cases, these eggs become a source of food for the community during a time of need.

"Guests" frequently live within the ant colonies. Some are very well mannered and even do the ant housekeeping. But other guests go about eating the ant eggs and destroying the community. And none of the home ants seem to mind! Once again, the chemicals given off by the "guest" ants must be highly prized or needed by the ants within the colony.

One of the more interesting ant guests is the spider. Ants have *six* legs. Spiders have *eight*. Yet, when spiders are guests in ant colonies, *the spiders walk on only six of their eight legs*. They seem to make believe their other two legs are *antennae*, the sense organs that stick out of the ants' heads. Are the spiders making believe they are ants to fool the others? Or do the spiders believe they have become ants?

In either case, we know very little about the chemicals and smells that are so important to ants. Scientists have hardly begun to identify them. But we do know—as seen in some of the very strange behavior of the army ants—that these chemicals play a very important part in the ants' lives.

Shortly, we will see just how strange and terrifying the army ants can be. First, let's take a closer look at their equally wondrous cousins.

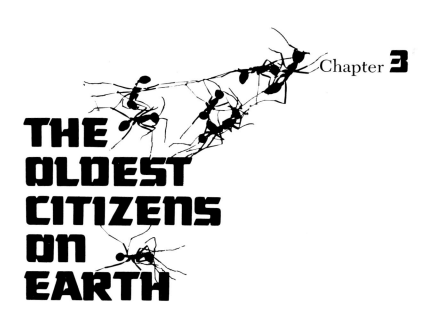

Chapter 3

THE OLDEST CITIZENS ON EARTH

You've probably often heard that truth is stranger than fiction. In the case of the ants, you will find this is certainly so. Ants are usually described as being large, small, red, black, yellow, green, blue, and combinations of each. There are easily as many different kinds of ants as there are words in this book—at least 6,000! As one biologist put it: "Should man and woman ever reach the stars, the ants will probably be right along with them!"

While the largest known ant measures only about two inches, the average ant is much less than an inch long. Believe it or not, the tiny ant has been around for more than 100 million years! The mighty dinosaur perished long ago. So much for large size always meaning strength! Humans have only been around for a million

Army ants tear into the flesh of a katydid many times their own size.

years or so. The ants seem to be the "senior citizens" on earth.

Ants have learned to survive almost everywhere on earth—in swamps, deserts, jungles, forests, city sidewalks, boats, airplanes, mountains, and coral reefs. About the only places ants have not conquered are the icebound Arctic and Antarctic areas where there are very cold temperatures and lack of food. Not only are ants ancient, numerous, and able to survive, they are also the most social animals.

Although writer Mark Twain considered ants "the dumbest of all creatures," scientists agree that ants have mental powers greater than most other insects. They seem to have memory and learn from experience. Experiments have shown that ants placed in a maze easily find and remember their way to food. Yet these same ants cannot work out even the simplest problem. For example, they build mounds of soil and move huge piles of earth. But the ants cannot figure out how to use the high mounds they build to reach food that is out of their reach.

Surprisingly, Queen ants can live to the ripe old age of 15, whereas workers only survive to

Actually, ants can retrace their way through a maze for food. But ants cannot solve other simple problems.

be 5 or 6 years old. An ant's nest may last 10, 20, 30 years, or even longer if it remains undisturbed. Compare this with the ordinary bee. Its hive (nest) lasts for only *one* season. Since ants can learn through experience and memory, the older an ant colony becomes, the better it is run.

The ant's body is also a kind of wonder. It has three main parts: the *head*, the *thorax*, and

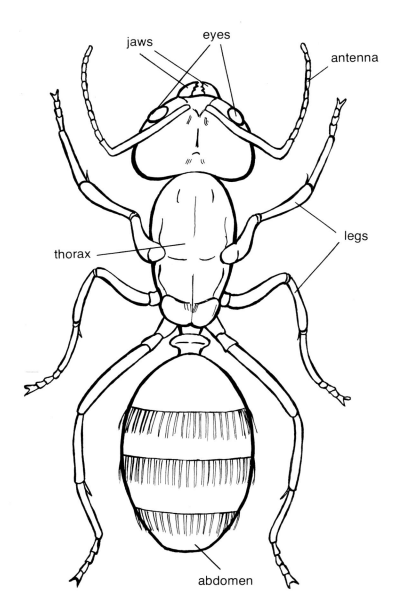

jaws eyes antenna legs thorax abdomen

Diagram of a worker ant.

the *abdomen*. Attached to the thorax, or middle section, of the ant are three pairs of legs. The abdomen contains the stinger and a device for releasing poison. Although most ants have large eyes on either side of the head, *they don't see*. Most of the ant's knowledge of the world is obtained through its amazing *antennae*. These sensors or feelers rise from the front of the ant's head.

While the ant is awake, the antennae are in constant motion. They can feel and smell objects from all sides. Would you like some small idea of what it must be like to live as an ant? Shut your eyes and cover your ears. Imagine two long rods sticking out of your head. These rods feel and smell for you. If you're walking down the road and come across another ant, you use your antennae to recognize your neighbor.

If the ant you have "seen" doesn't belong to your nest, you'll be able to tell by her smell. Chances are, you will then have to fight with the stranger. If so, prepare for a fight to the death. If she is one of your sister ants, you will probably "kiss" her. The "kiss" will be moving food from your mouth to hers. This food has been spit up from a place in the abdomen known as the *crop*, part of the digestive system. The crop is known as the "social stomach."

The social stomach, which is separated from the ant's "personal stomach" by a valve, is a very important key to the success of the ant. You see, unlike bees and wasps, ants don't usually have a storage space in their nests. Each ant carries its own food supply with it. Not only would it be expensive, both in material and labor to build storage cells for food, it would also be impractical. How would they move their stored food if they decided to set up housekeeping elsewhere?

Like knights in armor, ants wear their skeletons on the outside of their bodies. This hard envelope protects the soft, boneless body inside. The adult ant is born "fully grown," and it never grows any larger. Obviously, there are no weight problems in the ant colonies. It's impossible to get fat inside a permanently shaped, hard skin!

In any ant colony, there is one Queen. With rare exceptions, the Queen lays all the eggs in the community. At birth, the Queen ant has wings. But after she mates with a male, the Queen either loses her wings or breaks them off.

The ant has a hard protective coat around its soft body—"a knight in armor."

She's finished flying and won't need wings. Actually, the male ant won't be needed any longer either. So, she loses him, too. From then on, it's an *all-female* society—workers and bloodthirsty soldiers are all "sisters."

Chapter 4

A JOB FOR EVERYONE

You may have heard the word *metamorphosis*. It is a long word for one of the most amazing mysteries of all creation. Metamorphosis is the changing of an animal, as if by magic, from one form to another, as in the birth of an ant.

The ant begins life as an extremely tiny egg. If you turn over a rock and see a swarm of ants and little white bodies, you may think, incorrectly, that the white bodies are eggs. They're not. Eggs are much smaller. These little white things are *larvae*. They have emerged

28

Some worker ants act as nurses to the larvae.

from the eggs. Larvae are legless and quite help-less in most cases. Worker ants have to feed them either tiny morsels of solid food or liquid from their social stomachs.

In time, the larvae grow into *pupae*. Some ants even spin cocoons, just like caterpillars and silkworms. After a resting period, the ant slips out fully grown, almost ready to begin life as a worker, nurse, or soldier.

Some of the ants become *harvesters*. They are the farmers among ants. The harvesters ac-tually raise their own food by growing under-ground crops in complete darkness. Other ants become *herders* because they actually herd, pro-tect, and care for their miniature "cattle," (aphids) in addition to "milking" them. From them, the ants obtain a liquid that is not milk but equally nourishing.

There are other ants who become *thieves*. They steal food supplies from termites or even other ants. Strangely, the Queen of the thief ants is 5,000 to 7,000 times larger than one of her average workers. She is so large that she cannot possibly take care of her tiny young. So, wherever she goes, several small workers always go with her.

Amazon ants capture slaves to do the daily chores.

The *Amazon ants*, still another group, cannot live without slaves. Their bodies, brains, and social habits are specialized for battles. These battles provide them with the slaves they must capture in order to get their work done for them.

But the most terrifying ants in all the world are the notorious army ants. They make their "cousins" seem like playful pets.

Chapter 5

TERRORS OF THE TROPICS

Until recently, we have known few facts about the army ants, since most of them live underground and are difficult to study. But we've been learning more about these strange creatures. And the more we learn, the more mysteries we solve. Over the years, many stories have been told about their frightening behavior. Jungle movies frequently show scenes of people and

33

animals running for their lives to escape marching armies of ants. Accounts of these raids read like horror stories. But, in truth, many of these "accounts" are just more myths.

Although some ant armies will occasionally eat small animals like lizards and snakes, or perhaps even goats and horses, their *usual* diet consists of other insects—not larger animals.

In fact, there are many villages that greet the arrival of the army ants, not with fear, but with cheers and celebration. When the army ants arrive, the people get their homes thoroughly cleaned. The ants act as tiny exterminators. It's really quite simple. When the army ants move in, the villagers move out! Perhaps they vacation with friends, while the ants eat every living thing in sight—from cockroaches to field mice. In a few days, the army ants move on, and the villagers return to their clean homes.

Army ants are really two similar groups of ants—the *driver ants* of tropical Africa and Asia and the *legionary ants* of Central and North America. There are about 200 different kinds of army ants within these 2 large groups. They have colonies from a few hundred thousand to as

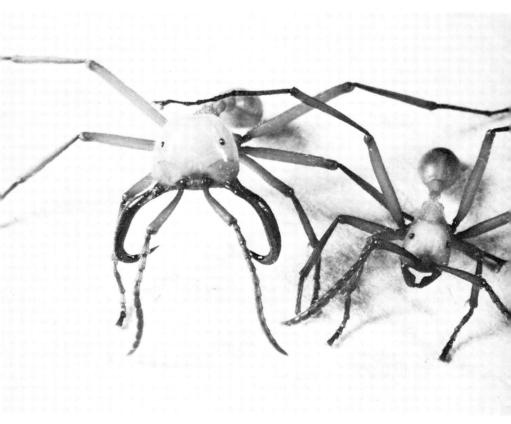

An army ant soldier (left) with a smaller worker.

many as 30 million individuals. They are *all* flesh eaters. All move their camps from place to place and get their food through periodic raiding parties. Since almost all the driver ants of Africa live underground, most of our information about army ants comes from the study of legionary ants. They live their lives on the ground.

Each colony of army ants consists of a Queen, males from time to time, and hundreds of thousands of workers. The smallest workers spend most of their time feeding the larvae and tending to the Queen. The medium-sized workers are the raiders, and the largest are the soldiers. Their sharp jaws protect the colony by tearing apart their enemies. The largest army ants might be 1¼ inches long, but most of them are much less than an inch.

Army ants are blind and wingless. Even the army ant Queen is wingless. There is one exception, however, and that is the male. He has wings and can see perfectly. The only problem is that he comes and goes so quickly, he's hardly any help to the colony.

Most other ants live in long-term homes— perhaps in ground tunnels or in trees. But army ants have no such home. They are forever wandering from place to place and forming nests with their own bodies. Each ant holds on to another ant's legs until a living rope is created. This chain swirls and winds around thousands and thousands of times, into one huge ball. This ball is the nest. While they're making the nest, one ant hooking on to another, many army ants

Dangling from a raised object, the ants create a living rope.

must support over a hundred times their own weight!

Deep within the nest of ants, fully protected from the outside world, is the Queen and her helpers. In cooler weather, the ants move closer together for warmth. In warmer weather, they spread farther apart, creating their own built-in air conditioning. The army ants can adapt to almost all conditions.

One of the great mysteries of the army ants is how their huge colonies function so much like a well-trained military unit. Remember—these insects can neither see nor hear. Who gives the orders? Who tells them to build their nests or to move out on a raiding party? Who decides when they should make camp or rest? Who selects those who will tend to the Queen and who will leave the nest on a scouting expedition?

It seems as if the army ant colonies are ruled, not by commands from their Queen, but by the amazing number of eggs she lays. Unlike other Queens, the army ant Queen lays from 10,000 to 100,000 eggs within a few days. Somehow, this unusual number of eggs she can lay seems to signal all the actions of the colony.

It's important to remember that all ants go through four stages of development—egg, larva, pupa, and adult. Studies have shown that whenever army ants are on the march, there is always a large population of hungry larvae in the colony. Not only that, but when these larvae start spinning their cocoons, about two and a half weeks later, the ants *stop* marching and make a

Thousands of army ants form a nest by holding on to one another, interlocking their legs.

camp. They stay there for about three weeks. At this point all army ant activities slow down and there are very few raids. That's because there are very few hungry larvae. The larvae have become pupae. Everything seems to come to a halt. The ants all seem calm, except for the Queen. She begins one of her fantastic egg-laying sprees.

Once a year, the Queen will lay about a half dozen Queen eggs and a few thousand winged, sighted male eggs. And then all kinds of exciting things happen. It appears to be very much like a political convention. Usually, the first Queen born becomes "elected." Mysteriously, some of the workers just start to follow her. She soon leaves the nest with a nice group of loyal followers, and she sets up separate housekeeping. For a day or two, the old and new army ant colonies remain in friendly contact, but then they split forever.

The unsuccessful Queens are abandoned and left to die. The males, gifted as they are with eyes and wings, fly off after mating. They are never heard from again. Perhaps they go in search of other Queens from other families. But the reports have not been as romantic. Many thousands have been seen clustered about street

After mating with the Queen the winged male ants are never heard from again.

lamps and headlights. They can never return to their nests. Most die because of lack of contact with an army.

Meanwhile, back at the army ant camps, larvae are popping out of their eggs, and strange things begin to happen. It seems as if the larvae give off some kind of exciting chemical smell to the workers around them. Whatever the chemical action, the larvae cause a great deal of activity in the nest. The nest becomes overcrowded. And at this point, the trouble starts.

The whole colony comes alive with motion and, before you know it, some poor souls are tossed out of the nest. These outcasts are the nervous, adventurous "scouts." They may not want to be the first out on a raiding party, but they're forced out to battle whether they like it or not.

Soon, the raiding columns are well set up. The workers stream out of the nest and follow chemical trails to the food the scouts have located. Once found, other workers run back to the nest with food for the hungry ants.

At the same time, another unusual thing is happening deep inside the army ant nest. The last batch of pupae are about to break out of their cocoons. If the smell of the larvae coming out of their eggs was exciting, apparently the smell of the young adults must be absolutely wonderful! The appearance of the young adults throws the nest into a frenzy of activity!

Raiding parties now rush back and forth with food for all the new ants. The raiding continues

Some unborn army ants in the pupal stage have no cocoons.

A grub is brought back to the nest by an army ant raiding party.

for several hours. Suddenly, the new ants start moving out of the nest to join the main raiding columns. They travel about 100 yards or so and stop. By that time, a new nest has started forming. The young ants return to the new nest.

Things don't really start to settle down for the army ant colony for about two weeks, when the new ants become adults. Maybe they lose their attractive smell. Regardless of why, the rest period begins again. The Queen again lays eggs. And then the whole cycle begins anew.

What makes the Queen lay "Queen eggs" and male eggs only once a year? Who tells the Queen that another year has passed? Why do some eggs develop into soldiers, while others remain small nursemaids? What is it that gets the workers excited and sends them off on hunting parties? No one really knows. Because they can only guess, scientists cannot figure out how to stop the raiding parties. They can't even figure out where the nests will be from month to month. The ways of the army ants remain a mystery.

Chapter 6

The army ants behave not only in strange and, to us, mysterious ways, but they appear bloodthirsty and vicious, don't they? After all, with such a highly organized and social culture, wouldn't it be wonderful if these insects could become even more like us humans? If they were more like us, they would show greater respect for each other and for the rest of the animal world. But, before you go too far with that sort of thinking, there are a few things you should consider.

Face to face with the army ant: the terror of the tropics.

A very long time ago, ancient civilizations used the heads of ants to help heal wounds. First the ants would bite the wound closed. Then their bodies would be chopped off. Only the head would remain clamped on the wound as though stitched by a surgeon.

Some not-so-ancient civilizations tested the strength and courage of young men and women about to be married. They would put these people in ant cages and observe how they reacted when bitten.

A very modern civilization took great pleasure in what is known as "ant torture." Enemies were tied to army ant nests during the hunting and raiding season. Hundreds of thousands of killer ants would tear the enemy to shreds. It was a horrible sight.

Now, whose behavior is more terrifying—the ants' or the humans'?